2020 Family Bill Overview

Sustaining our Future with Authentic Family Law

Published by Clean Law 2019

International Standard Book Number: 978-0-9975507-8-8
International Standard eBook Number: 978-0-9975507-9-5

1. Family 2. Science 3. Law
4. Education 5. Social Science

Printed in the United States of America.

All rights reserved. Except as permitted under U.S. Copyright Act of 1976, no part of this publication may be reproduced, distributed, or transmitted in any form or by any means, or stored in a database or retrieval system, without the prior written permission of the publisher.

For more information:
Clean Law
www.cleanlaw.today

www.cleanlaw.today

PAINFUL DISCONNECTIONS

Traveling through life we learn that when a national problem like AIDS breaks out, then there is no humanity in various governmental disconnects. Our only chance for efficient solutions are for either the private sector to solve the problem and save lives on a massive scale, or else for the federal government step in and eventually intact a bill that has the appropriate bridges and child safety nets.

By Aaron W. Wemple

CONTENTS

Preface   7

Introduction   9

1. The Littlest Grapes. 11
2. 2020 Family Bill Overview. 13
3. The Unsustainable Freak Show 29
4. Truth Hubs for Sustainable Families. . . . 35
5. Child Cannibalizing Contracts 37
6. Shame & Professional Development. 39
7. The New Family Law 2.0 43
8. The Emergence of Family Sustainability Leadership 45

Preface

Hello world!

Do you like new things?

I think it's in our DNA. We like new things. I could have ten ideas and you would have ten new ideas about my old ideas. And vice versa. Thinking brand new is really in our nature. But buying into it, well, now that's a different story.

It's hard to go out on a limb, so to speak. It's hard to change. For example, if I'm a boxer and that's what I've always known to win, then why put my hands together like in this picture of holding a duckling when that would feel uncomfortable to me? You're habits tell you to swing but you know you should hold. Changing our behavior takes effort.

Cardinals play the Cubs, Rams play the Bears, Republicans versus Democrats, Plaintiff versus Defendant, divisive mechanisms seem to be all we know. But the classical contingency traps we lay for ourselves in these old "versus" games leave out crucial elements. Where's the newness? Where's the kids? If one side is always the winner and the other is always the looser, then why can't kids fit in?

Whether these competitions are in family law making deals or in national policymaking deals, it's like wearing old shoes that just feel comfortable. Why would we pivot even an inch to try something new?

What if policy makers could literally see evidence and files from the nuclear families perspective? When we supply policy with the preventative measures inherently found in child safety files, then tomorrow's policy will incorporate family sustainability automatically and help prevent divorce.

What if policy makers never get to weigh decisions from the children's perspective and always only feels the incredible powers of lobbying perspectives?

How many children out their wish that they could get their parents back together? System Up helps those dreams come true. Saving their first loves of both of their parents like fine gold. We'd no longer be sinking them right from the start in the quicksand of the alternative system.

Introduction

In the above scenario, where two disconnected parents are trying their best to explain to their children that an idolized show is not what it claims, it sounds unsustainable and self-defeating doesn't it? Like blocking the children out from what its intentions obviously are. What if a competitive show like the World of Dance never judged participants on dancing skills? That would be a self-defeating purpose like family law.

In the future, families would be more sustainable by proving skills like parenting. Not by gaming the mechanical disconnects of the family law system. Just like dancers get to prove their dancing skills on the World of Dance and singers are allowed to prove their singing skills on the Voice. All three shows will be sustainable when families are allowed to family. When parents can prove their parenting skills.

You see, unfortunately, the classical family law disconnecting dynamics are unsustainable at best. Self-defeating at worse. And maximum dependence on double self-defeating disconnected dynamics for children who are sentenced to that machine. Like two boxers pulverizing each other in a boxing ring, that practice is punitive, adversarial, misleading, and exhaustive. For the littlest grapes in these big conflicts, they are bound to those double financial depletions, double emotional depletions, double physical depletions and more. Child cannibalizing deal making. Which is kind of old school and barbaric if you think about it. Two parties free to fight with every ounce of everything until one or the other is pulverized. What message does this send to children?

What if builders who build buildings never had to prove building skills? What if car brake manufacturers never had to prove their braking abilities? What if food and water producers never had to prove that their products were safe and consumable?

Perhaps it's time to inflate the mandate. To start balancing the little grapes equal to those big conflicts. What if we valued children and child safety files as much as we do adults and adult fighting files?

The 2020 Family Bill would finally allow <u>all</u> families to be sustainable. It pivots those unsustainable contract making ways to child safety ways. It's a seamless deep-level integration of state-of-the-art scientific law with classical family law. This preserves the integrity of other fields of law. And it opens a brand new path towards crime and child abuse prevention, and a more sustainable future for us all. Just like a world of families!

Trials can be thought of like a sports injury. There's rehab after the injury and prehab before the injury. Scientific law is like prehab, or prevention. When engineers use scientific laws to build things up like skyscrapers and bridges, then they know ahead of time whether they'll stand or fall. Today we can integrate scientific law with classical family law beginning with child safety files. And when we integrate these safety files with adult fighting files, then we eventually build kids up in every possible way while their families are being torn down. Thus, balancing the little grapes equal to those big conflicts. That's sustaining our future!

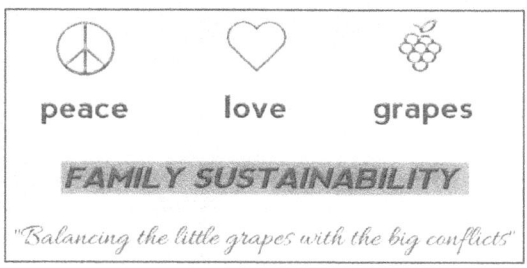

1. The Littlest Grapes

The littlest grapes of those biggest conflicts are the most fragile indeed. They get deflated and feel more pressures more easily than big grapes do. In fact, big grapes may not even notice the pressures that little grapes endure. My child, for example, cried when he thought his mom did not know what to cook him for supper. I, on the other hand, would have just ate anything. The littlest things can hurt. So, balancing these pressing scales for the little grapes are necessary. In fact, it's socially urgent.

Studies show what experience knows. That children struggle to understand the monumental conflicts like divorce, abuse, policy making, and even bullyisms that may go through. They find it hard to talk about those issues without getting hurt even more. Often they don't have a vocabulary to explain what they're feeling. And most children have barriers to adjusting. Some never adjust at all. And sadly, we've all heard of the young children who have taken their own lives due to bullying. We don't only know these sad facts from education. We know them from experience.

But now there's hope. Now there's good, quality bonding opportunities for our families to sustain. To keep the little grapes from getting pressed and sour by big levers too hard. Even a bad day can now feel safer. But it may take an act of Congress to go as far as it needs to go.

Those closest to the little grapes are their first responders when conflicts happen. And that area has an unmet need in society. IKEA published a very impactful experiment in 2018 where they placed two plants inside of a school in a common area. A recording loop was played to one plant talking deconstructively to it for 30 days. And a recording loop was played talking constructively to the other plant for 30 days. The one plant that absorbed all those deconstructive words for 30 days wilted. But the plant that felt all those constructive words for 30 days **flourished**.

Children in classical family law disconnects are being ignored, dismissed, and unrepresented. Children of abuse and neglect need help too. And this holds true in policy making because children can't afford lobbyists and they can't stand up as activists. Little grapes need balance there too.

But besides praying for them and having faith with them, what else can we do to sustain our little unsustanable families?

Well, now we can literally balance the scales for our littlest grapes of the biggest conflicts. An immediate parent or immediate connection can give them the sustainability that they need and deserve. A family sustainability author gives them what other systems can't give them. An opportunity to flourish on record while not flourishing on record. Every organism needs a positive and negative balance to survive. And if it works for plants, then how much better will it work with law?

"Children are great imitators, so give them something great to imitate!"

**Studies show that people are 40% more likely to remember something if they write it down.
The future is 100% more likely to remember something we write, file, and save. Plus, there's an overabundance of negative pens and records out there wilting us all.**

Like Adult Fighting Files

Like Child Safety Files

2. 2020 Family Bill Overview

The Family Sustainability Initiative

Family sustainability can be thought of as being like crop sustainability. When the Farm Bill was passed, that seemed to be a natural, fitting, and healthful issue that everyone could agree on since we all need to eat. According to Farm Policy Facts, conservation was the primary focus of the Farm Bill. The goal of sustainable agriculture is to meet society's food needs today without compromising the ability of future generations to meet their own needs. The 2020 Family Bill is like a family taking care of itself without compromising that ability for next-generation of families.

If we didn't know to take care of the soil and the crops, then we could all eventually perish from that lack of knowledge. And we don't let the IRS, for example, watch over the Farm Bill. That would be a world of hurt. We have specialists for watching over the Farm Bill and crop sustainability. Congress mandates the United States Department of Agricultural to oversee the Farm Bill policy.

Like many of us, for twenty years I went through family law practices and through family law battles trying to prove what a great parent I was. And for twenty years they locked me up, pushed me down, set me back, and took away my grapes trying to prove that I was at the wrong show. And there was no place to file those constructive records. After all, they said, isn't it normal for good parenting skills to be judged in the same department that judges the world's worse criminals? They laughed and said "What do you think this is, the world of parents?"

Before we get into the 2020 Family Bill Overview, let's think about self-defeating industries. If a dancing show never judged dancing, a singing show never ruled on singing, and a talent show never made decisions on or filed anything about talent, then wouldn't those industries be self-defeating? Likewise, no one judges or files family sustainability.

We can never claim that we're helping families if at the same time we're allowing them to be pushed into disconnection sentences like dropping a grapes into a blender. Of course, we're making a mess!

Without a Family Sustainability Plan

"An unraveling love affair is not grounds to allow truly innocent little grapes to fall into more unsustainable pulverizing pressures harder, faster, and longer than criminals are allowed to fall."

This is our little grape's souring subjected to the classical family law practices alone:

PRESSURE²

(Without the 2020 Family Bill or Family Sustainability Plan)

#ExitTheChildren

It really is time to change the climate for little children in one of the very few remaining ecosystems without child safety measures. It's time to redeem the next generation. It's time to re-regulate the divorce court industry. It's time for the 2020 Family Bill to allow Family Law 2.0 with child safety files on a national level.

With a Family Sustainability Plan

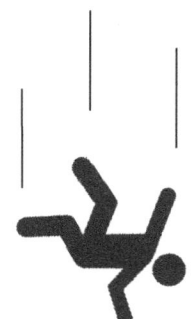

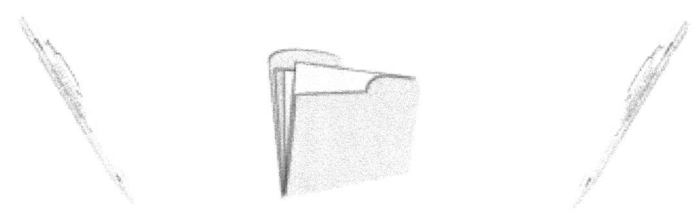

2020 Family Bill

With Safety Files

These are tomorrow's sweet and voluptuous little grapes revived from those pressures from state-of-the-art Family Law 2.0 with child safety files:

RELIEF²

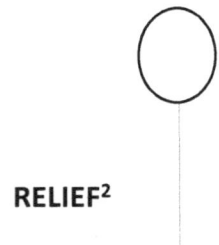

(With the 2020 Family Bill or Family Sustainability Plan)

#SavedTheChildren

World of Family

There's a famous quote that goes, "Do people gather grapes from thorn bushes or figs from thistles?" And the answer is a resounding NO! People do not gather grapes from thorn bushes. That's like taking someone snipe hunting. Like family law. Those mechanisms used for criminals should in no way, shape, or form be the same mechanisms used for families. And the pitfall of calling divorce law "family" law is wrong and insatiable.

You see, when two teams battle on the show called the World of Dance, then two teams of dancers are judged by other dancers according to dance categories like performance and technique. There are different classes like junior and team. The best dancers move on to the next round while the not so good dancers go home and keep on practicing. And whether they "win" or "loose," it's all about the dancing!

Family law decisions need to be more like the World of Dance. If both dance teams on the World of Dance battled in court and had to wear court clothes, speak court language, and move according to court mechanisms, then would they perform as well? Of course not. Would that be sustainable for dancing or unsustainable for dancing? That would be unsustainable. So, is that sustainable for families or unsustainable for families?

Industries change for a reason. Apple did not deconstruct the music industry. It made it better for the consumer. Netflix did not dismantle Blockbuster. It made the experience better for customers. Family law and policy improvements are not deconstructing the legal industry, they are making things better for the end users.

So in practice, what would a sustainable family competition be like? Would constructive evidence come out of the sadness of divorce? Would uplifting forms come out of broken hearts? Would a life system rescue the littlest grapes from the biggest conflicts in the not so living system?

Well, let's think of it in terms of overcompensating. In other words, what would a flourishing family mill look like? When we bring to the table just the constructive evidence, then couldn't the best parenting be a decision? Of course it could. Like the best dancers be allowed to continue to dance on the show while the not so good dancers are still be encouraged to dance. Those flourishing mills are a win-win for competitors.

What if that mill could be recorded to help every rule? What if those open-ended pitfalls could literally pivot into child safety nets, those mechanisms could

literally pivot into buffers, and children could literally
pivot into being formally built up? I mean, what if dancers could literally dance, singers could literally sing, and parents could literally parent? That's like the World of Dance, the Voice, and the 2020 Family Bill.

When we practice uplifting forms and demoting forms together, then wouldn't hearts be more resilient? What if this could help every broken heart heal?

When a life system inflates the littlest grapes while a differing system presses them from all directions by proxy, then wouldn't family sustainability be the result? Scientifically, it would.

And what if the goal of the 2020 Family Bill was to make family court like the World of Dance? Where the best performers won the show? The best love for their child gets to grow? And the not so good performers go home and practice for the next show? In those cases, even the not so good parent inherently gets better.

Family is as family does. Another famous quote goes, "A grape for a grape, and a fig for a fig." (Well, I just now kind of made that one up). But we really have to wonder what a family flourishing mill would be like. Wouldn't that be exciting! What a family flourishing mill in sync with the family erosion mill would be like for family sustainability? Couldn't it finally be like the "World of Parents?!"

The Unsustainable Family

On one the hand, if we bridle the dancers on the World of Dance and hold them back, then is that the best way to develop dancers for dancing? What if we bridled the singers on the Voice with different pitfalls and mechanism during their battles? Would that be best for singers and their families?

If classical family law practices really cared about the best interests of children, then wouldn't they ask families to compete and prove who makes the healthiest meals, who teach their child the most, or who offers the safest living conditions? Or, at least publish out guidelines like the Office of Disease Prevention and Health Promotion publishes dietary guidelines?

Is the classical world of "family" divorce law unsustainable?

Well, it's safe to say that many of us know what a family erosion mill feels like. Even if it is universally painted as something else. Many of us know the deconstructive evidence that is placed on top of sadness. Either from when we were a child or as an adult. And many of us have filled out those demoting forms that break the broken-hearted. We all pay taxes for a different system pushing us into the grave quicker than those who do deserve it. All because we believe in a different department the way it is painted. We believe in a grape for a grape as advertised. But in classical "family" law, families are unsustainable.

When we engage adult fighting with child safety:

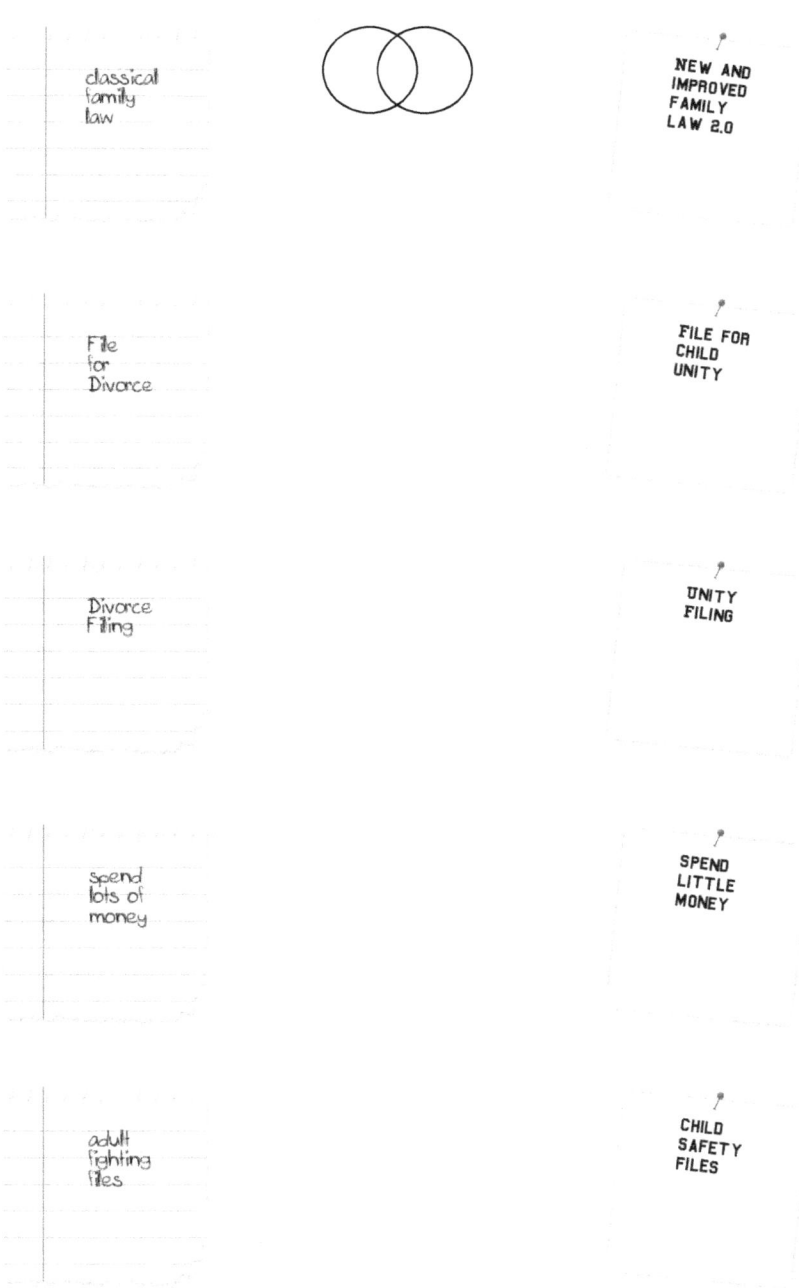

classical family law		NEW AND IMPROVED FAMILY LAW 2.0
File for Divorce		FILE FOR CHILD UNITY
Divorce Filing		UNITY FILING
spend lots of money		SPEND LITTLE MONEY
adult fighting files		CHILD SAFETY FILES

When we equally add family sustainability:

motion to continue	MOTION TO STAY CONNECTED WITH CHILDREN
family deconstruction	FAMILY CONSTRUCTION
"Mediator" divorce law instruments	CHILD-FRIENDLIER EYE-WITNESS INSTRUMENTS
Children have no rights to a voice and only family deconstructive evidence is shown	CHILDREN HAVE RIGHTS TO A VOICE AND TO SHOW CONSTRUCTIVE EVIDENCE
code kids out	CODE KIDS IN

"Balancing the littlest grapes equal to the biggest conflicts"

When we finally add Family Law 2.0 with child safety files:

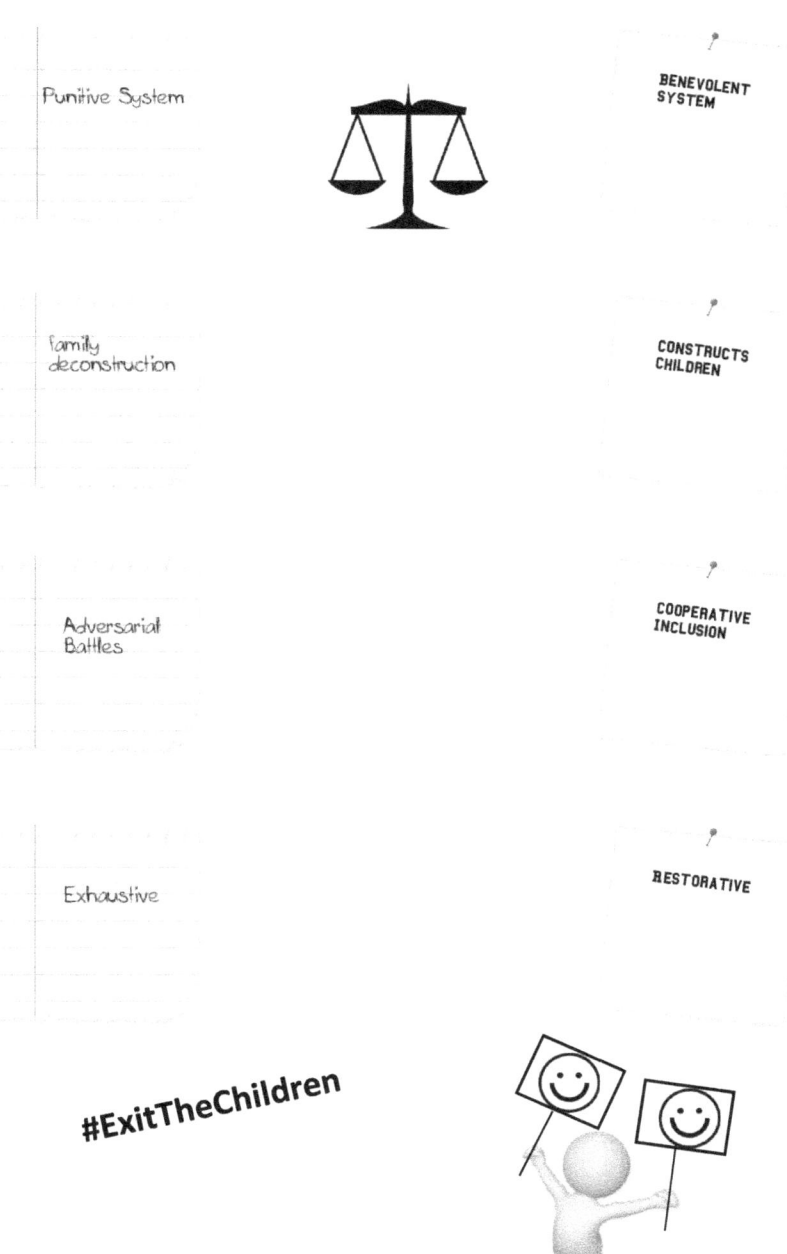

Punitive System	BENEVOLENT SYSTEM
Family deconstruction	CONSTRUCTS CHILDREN
Adversarial Battles	COOPERATIVE INCLUSION
Exhaustive	RESTORATIVE

#ExitTheChildren

"Balancing the littlest grapes equal to the biggest conflicts"

3. THE UNSUSTAINABLE FREAK SHOW

"Insane Inequality"

The Ugly Drain

Sub-criminal divorces are way more off-balance and polluted than we think. In California, ABC 10's Luciano, L. (2018) says that those with the basic knowledge of criminal law may walk along with family law practices for years and years fully expecting the right to an attorney someday, transcripts, the right to a jury, investigations to track evidence to prove a claim, the right to bail when accusations are made, the presumption of innocence, a formal voice for children, and a formal place to document the effects such as in table, chair, pen, paper, forms, and mechanisms for the children's safety during these family wrecks. The sad fact is that family law clients are not equal to criminal law clients. There are severe oversights. And that pains us all. In fact, there are at least thirteen inequality walls between criminals in criminal law and children in family law. And a recent study from a scientific perspective reveals that there are 10 major structural voids in classical family law makings which children are chronically falling through like draining the ugliest ducklings.

Child safety has historically always been forgotten about until way after the fact. Take child car seats for example. The automobile was invented in the 1920's. But Good Housekeeping Magazine Reporter Smith (2018) says that the first child safety seat with a seat belt wasn't invented until 1964. That's almost 50 years of child endangerment people accepted. Are the children of divorce freaks? Are they in danger while the rest of us accept it?

Sadly, the law of the land seems to find serial killers like Ted Bundy much more valuable than truly innocent family members. To be a serial killer in criminal law and criminal court is beautiful compared to being an innocent child in family law and family court. If life were levels of a building, then criminals are given valet parking and taken up to the thirteenth floor while innocent young children can't even pay for parking and are taken to the drain in the basement. That's insane inequality. A freakish oversight.

#ExitTheChildren   RELIEF[2]

What if teachers treated one group of students thirteen times more important than another group of students day in and day out? What if teachers provided the favored students with homework assistance, a way to prove answers, a table to work on, a chair to sit on, pencils to write with, paper to write on,, and a folder for their work, but did not provide those same necessities to the unfavored students? What if we paid taxes for thirteen times more safety features in a luxury car for criminals than for children? Maybe it's time to reimagine child protection?

Luciano (2018) says that those beautiful swan features in criminal law make sense because of the high-stakes of prison. But the lack of safety features in family law doesn't make sense because the consequences are much more than losing your freedom. Especially for those children who are uniquely positioned to feel the impacts, pains, and ugliness of both sides at once. And they deserve to be the safest, not the most endangered, when their families wreck.

Yes, the absence of child safety creates the ugliest ducklings and life drains. In science we think of these as heat sinks, or the ground for electrical circuits. It's where all the negative energy goes.

Parker (n.d.) says that many of the parents involved in that swamp say that they would rather spend years in prison or else die than lose connections with their children and property. And many have. Can you imagine what those children who feel twice that pain would say?

Plugging the Drain - Balancing the Dirty Divide

Judges often appoint special advocates for the children of crimes to give them a voice in court. But judges refuse to appoint special advocates for the children of truly innocent parents. What's wrong with these ducks? Is their only worth to be drained?

All children are valuable. The children of truly innocent families are worth as much as criminals even if we don't prove it.

Aspiring to be a criminal is, unfortunately, not just a step up in society that these children can feel while they are in fear during those wrecks. But it's at least thirteen steps up. Those children grow up and someday realize that they could have thirteen times more safety with the authors of their life if they were serial killers instead of children of divorce. That's the most formal message they get. In their world, it's literally a promotion times thirteen to chose a life of crime. It's time to reimagine child protection.

Parker, W. (2019) reports on the statistics for the children of divorce. The physical and emotional effects are staggering. Children suffering in the ugliest ducklings drain are more likely to experience injury, asthma, headaches, and speech impediments than children whose parents remain married. Teenagers in single-parent homes and in blended families are 300 percent more likely to need psychological care than teens from intact families. Children in the ugliest ducklings drain may have more psychological problems than children who lost a parent to death. And when you understand things from a scientific perspective, then it's easy to understand why. It's easy to elevate child safety.

Plugging the Drain - Balancing the Dirty Divide

Judges often appoint special advocates for the children of crimes to give them a voice in court. But judges refuse to appoint special advocates for the children of truly innocent parents. What's wrong with these ducks? Is their only worth to be drained?

All children are valuable. The children of truly innocent families are worth as much as criminals even if we don't prove it.

Aspiring to be a criminal is, unfortunately, not just a step up in society that these children can feel while they are in fear during those wrecks. But it's at least thirteen steps up. Those children grow up and someday realize that they could have thirteen times more safety with the authors of their life if they were serial killers instead of children of divorce. That's the most formal message they get. In their world, it's literally a promotion times thirteen to chose a life of crime. It's time to reimagine child protection.

Parker, W. (2019) reports on the statistics for the children of divorce. The physical and emotional effects are staggering. Children suffering in the ugliest ducklings drain are more likely to experience injury, asthma, headaches, and speech impediments than children whose parents remain married. Teenagers in single-parent homes and in blended families are 300 percent more likely to need psychological care than teens from intact families. Children in the ugliest ducklings drain may have more psychological problems than children who lost a parent to death. And when you understand things from a scientific perspective, then it's easy to understand why. It's easy to elevate child safety.

A New, Child-friendlier Pond

In Decatur, Illinois, Macon County is also known as "make-up" county. But now, police, caseworkers, parents, families, and friends have a brand new pond beautifying those children of divorce, abuse, neglect, and those who've just had a bad day. This brand new pond has escaped out of necessity and set some sinking captives free. A place where constructive scientific law meets classical family law. And they work together side-by-side seamlessly. Mississippi College of Law Professor Alina Ng Boyte (personal communication, December 5, 2018) calls Family Law 2.0, "...an innovative new program mitigating the effects of divorce on children." And Alina says that "The family law system has its flaws and children are the most innocent ones who pay the price for those flaws. Even with the CPS and abuse/neglect cases."

The vision is clear: Clean Law.

The goal is simple: "Reimagine child protection."

The reason: Because everyone deserves a first chance.

Clean Law achieves local and national value in family sustainabilities through innovative policies, practices and inclusions which empower next-generation safety to its fullest.

An adventure of discovering new life.

Sources:

Luciano, L. (2018, May 10). The problem with family court. Retrieved February 19, 2019, from https://www.abc10.com/.../the-problem-with-fami.../103-550687204

Parker, W. (n.d.). Statistics About Divorce and the Impact It Can Have on Children. Retrieved March 30, 2019, from https://www.verywellfamily.com/children-of-divorce-in-america-statistics-1270390

Smith, L. (2018, November 05). The Evolution of Kids Car Seats. Retrieved January 19, 2019, from https://www.goodhousekeeping.com/life/parenting/g2870/car-seat-history/

Wemple, A. (2018, October 28). Family Law 2.0. Retrieved September 5, 2018, from https://iprlicense.com/Books?SelectedPageSize=20&OrderBy=1&MainSearchBar=family law 2.0 aaron wemple&BooksFilter=1&HasRights=1

CHILDREN ARE

ALWAYS FRAGILE

4. Truth Hubs
for Sustainable Families

The Coming Age for Truth Hubs and Family Sustainability

In years past, the disconnected always tended to disconnect the disconnect. But the connections that are finally enabled by Clean Law's child safety files (Family Law 2.0) and the 2020 Family Bill are the only automated **truth hubs**. Every other interaction, or disconnect, even if they say "we're with you " or "connection," are in fact not truth hubs. And they often have self-absorbing motive.

There's only one truth hub between every parent and child. That's what Clean Law and the 2020 Family Bill maximizes for family sustainability and next-generation safety. Everything else, every other word, every other report, interaction, file, or system is <u>not a truth hub</u>. In fact, their hub might be financially prohibitive while yours are inhibited. But make no mistake about it, only yours is right and the most valuable. Like when you order a pizza and its always exactly as you ordered versus ordering a pizza and its never what you ordered, but always what someone else wanted instead.

The Family Sustainability plan is for truth hubs and next-generation safety to flourish as a legitimate practiced. And the community health and welfare will increase while community lack of health and lack of welfare will decrease. A proof of concept already exists in one community. In fact, it turns intellectual bob wire into intellectual networking wire. Which means that family sustainability is on the verge of being accepted as a scientific law.

Practicing truth hubs to balance the practices and even games of the disconnected makes since for sustaining families. The ways of practicing truth hubs are un-leverageable. Whereas the other hubs have two times all the leverage in the worlds if its needs be. How wasteful.

4. FAMILY CANNIBALIZING CONTRACTS

Sadly, there are such things as family cannibalizing contracts. Contracts made at one point in time from overriding conditions which change against our will. Forcing a breech of contract and their consequences.

In classical family law, statistics like early death show that these family have cannibalizing terms. Those families who have experienced this can testify to how it feels. Those contracts, fighting files, and old legal dynamics can even cannibalize the vulnerable children involved with divorce. Any one can test this or document those consequences over the lifetime of a deal.

Take child support for example. Children are normally contracted to received a fixed amount of money each period based on the non-custodial parent's income at the moment in time which it was written. When it's written in terms of a fixed rate, then it cannot vary when income or expenses change. Yet, state law mandates child support to be a percentage rate. This means that every pay period it should be a percentage of that particular pay day. Using a fixed number instead of a built-in percent of income, like employers do to calculate taxes, can cannibalizes those children's support. Because each and every time that income changes, then that non-custodial parent violates the law. They can't follow the percentage rate mandate when they are locked into a fixed amount. Which again, over time, cannibalizes the support that those children need to survive and really do deserve.

Classical family law contracting, adult fighting files, and those strictly adversarial, punitive, and exhaustive dynamics actually render families unsustainable. Contracts with no reality-based, built-in remedies that trend with reality are contracts which cannibalize those families.

6. Shame & Professional Development

"Excluded remedies vs. built-in remedies"

As we know, appearance is not everything. Some things look inviting, but are really not. Contracts and courthouses can appear inviting, but they really may not be. Authentic child safety must include built-in contractual remedies. For both ends of those deals. Because like a fish out of water, so are the authors of our lives kicking the feet right out from underneath of them.

Unfortunately, there is a lot of money to be made in the future by excluding built-in contractual remedies. For example, if a term in a contract is a fixed amount not able to track with reality instead of a variable percentage rate that's able to track with reality, then those deals need modified every time that income's change or else one of the signers violates the law.

Sadly, there are child cannibalizing contracts of both sides of those deals. For example, if a custodian parent isn't receiving enough money to pay for children' food, clothes, and living expenses, then those contracts need modified every time expenses change. Yes, terms of contracts without built-in remedies can starve young lives. That's unauthentic "best interest."

Authentic child safety contracts including built-in remedies are not rocket science, or some new and novel idea. Businesses and bankers use built-in remedies in contracts all the time. The sad fact is that there's just too much money being made by making up and then "fixing" any ole disconnect when nobody is opposing it, and there's no "staying connected" police.

Employers deduct a percentage of our gross wages each and every pay period for taxes, IRS's, health insurance, etc. That's an example of a built-in remedy. Likewise, authentic prevention of "dead beat" child support is somehow deducting the statutory percentage each and every pay period. That would take out the jagged edges. The very draining jagged edges.

Bankers and corporations use built-in remedies all the time. Late fees for a missed payment, for example, is a built-in remedy. Keeping record of how much food, clothes, and living expenses that children need and then charging that amount plus some overhead is another example of a built-in remedy. But how in the world would deadbeat contracts ever get eyes to those financial accounting accuracies? Excluding built-in remedies is no longer the responsible thing to do. It is not financial legal in any world but family law to intentional harm little children.

Like ducks willed down the drain blind-folded and forced to swim upwards through a maze of revolving drain plugs to get to a safe and comfortable place with its baby ducklings, so are the many disconnects between accepted or not filings, vertical position tops, and horizontal positional flops. We are unsustainable little ducks thanks to those diabolical disconnects.

As you know, appearances aren't' everything. There's no humanity in disconnects. Remedies that are built into the terms of contracts helps both ends meet. Whether it's a poor mom having troubles paying for the bare necessities while raising a child. An injured dad having troubles paying for child support. Or, any other party locked into a deal and circumstances change. Built-in remedies can track with reality and never but their subjects in violations of the law and reality. There should be no money made from disconnecting others from their ability to track with reality. There should be no extra money made from disconnecting people from their ability to follow the law.

Like a fish in water, so could authors keep the living feet underneath of us all forever. Authentic safety, as you can imagine, helps people up to those disconnects, and then makes the paths safer for the next travelers. Especially for those children who are involved with those disconnects. Like lily pads willed to float through life with transparency, so would others be helped along the way to a safe and comfortable place with other lily pads. Because many connections between files would be accepted a little or else accepted a lot with no vertical differences and no horizontal differences because of those considerate connections. It's a family sustainable pond thanks to those considerate connections.

Judicial insecurities are real. Judicial insecurities like these are like food insecurities on steroids. Judicial insecurities coupled with disconnection insecurities breed a like of trauma. We should all be ashamed of ourselves. But we could do a billion times better for the next-generation

When we uncover all of the draining pitfalls in front of these ugliest ducklings, then we will create some really authentically beautiful swans. We may even have a pond full of swans.

Let's face it, lawyers, for the most part, who influence and write these contracts, are entrepreneurs. They may not think there's a need for personal and professional development which big companies provide. While the myth is that an entrepreneurial lawyer is ultimately justice, the reality is that lawyers need help to develop professionally.

Matveeva (2019) says in Forbes magazine that without feedback and development that any good company gives employees, entrepreneurs can be left out of progress and never understand what they are doing wrong. In study of 212 startups, Harvard Business School academic Noam Wasserman discovered that more than half of Founder-CEOs were still CEO of those businesses after three years.

Matveeva (2019) reports that this change is because entrepreneurs may not have the where with all to include a process which comes with scaling. And without a feedback process to maximize scaling, then a business very well could always be turning left, for example, when its clients need it to sometimes turn right.

Another oversight, she says, is the lack of development in leadership. How can anyone solve these oversights without a boss or a corporate professional development program?

Matveeva (2019) recommends a solution to solve the start-up problems lawyers face: recognize there's an issue and make a plan to work on it. Many lawyers should be recognizing this by now with all the feedback about the way they're handling things, and chat with each other about investing in development. Perhaps by getting a personal development coach.

Professional development leads to life changes. Professional development for the greater good leads to life changes plus added respect. The better that we're developed to do our profession, then the more revenue our company makes, the more satisfied we are with out job, and the happier our customers become. A smart and safe operation to get the help we need for jobs well done.

Getting feedback, a professional development coach, or a model corporate development program all work. Matveeva spoke to her coach Harriet Minter on tips for taking development to the next level. She said that it really is part of the job. In other words, it goes with the territory. "Part of your work is being responsible for your team, and you (as an entrepreneur) are your team." The first step is to pivot your mind and know that development is part of your job. Matveeva explains that the growth loop which follows feedback, learning, growth, practice, then more feedback, more learning, more growth, more practice is all fulfillment in of itself.

Reference: Matveeva, S. (2019, May 05). Professional Development For Entrepreneurs: Why It Matters And How To Do It. Retrieved May 7, 2019, from
https://www.forbes.com/sites/sophiamatveeva/2019/05/05/professional-development-for-entrepreneurs-why-it-matters-and-how-to-do-it/#76f8548f1f08

The essence...

...of being free...

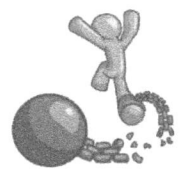

(Source: https:steetmit.com, 2019)

...is discovering new things!

7. The New Family Law 2.0

Coming soon . . .

"Woven Plush"

www.cleanlaw.today

8. The Emergence of Family Sustainability Leadership

In this chapter, we look at issues and strategies for developing leading practices of family sustainability in organizations. Three questions will help guide our discussion: How are practices of family sustainability being developed and maintained in organizations? What management styles in particular are involved with developing practices of family sustainability in organizations? How might these perspectives be fostered by other organizational leaders?

This approach draws from literature's in family, social-emotional learning and corporate social responsibility to describe the nature of family sustainability, the intents and approaches of organizations developing it, as well as their challenges. Case studies are drawn from small businesses and are examined to explore successful strategies of developing practices in family sustainability. These examples are analyzed from an organizational leadership perspective through management applications.

With this newfound interest in corporate family sustainability, organizations are seeking to develop practices and policies that are more family sustainable and socially responsible (BSR, 2006). Yet despite prevalent business usage of 'sustainabile' marketing, critics have argued that the overall impact has been unremarkable in achieving real goals of family sustainable practices within organizations (Daley & Cobb, 1989).

This calls for the overall discussion on this blog: How are practices of social responsibility and family sustainability developed and maintained by organizational leaders? The focus is upon various leadership perspectives. Examples are based on studies of small businesses, which tend to be vulnerable to volatile markets and bigger competition, and must struggle to find and implement easy sustainable practices when profit margins may be small and survival is closely linked to the bottom line.

The discussion begins by clarifying the concept of 'family sustainability' and outlining challenges encountered by organizational leaders trying to implement family sustainability practices as well as to increase social responsibility.

What is 'Family Sustainability'?

Sustainability has come to represent everything from environmental protection to supply chains. In this book 'family sustainability' refers to both social responsibility, ranging from issues of human rights and generational connectivity to personal families. Working from the literature of ecological sustainability, one principle of family sustainability suggests the interwoven considerations for organizational leadership empowering practices to shift neglected and unsafe vulnerabilities (Todd & Todd, 1994). Which, in turn, strives to reduce the negative influence of those footprints on the interdependent webs of society. Thus, helping to ensure meaningful paid work experiences in respectable conditions while contributing to the increase of employee morale and growth of healthy urban communities.

Family Sustainable Practices

An international movement that echoes this principle, and under which many organizational efforts have organized themselves, is 'corporate social responsibility' (CSR). Although the issues of CSR overall tend to be broader than the concepts of family sustainability adopted for this discussion, the CSR strategy is worth examining for organizational practices. CSR has been defined as 'treating the stakeholders of the firm ethically or in a responsible manner' (Hopkins, 2003, p.1) by recognizing a 'triple bottom line' of these stakeholders that includes people (employees, customers, competitors, communities, national, global), the natural environment, and investors. Authors for CSR practices tend to address a wide range of categories including environmental sustainability, employees' rights, suppliers, customers' rights, transparent and honest accountability, legal and honest operations, and global citizenship
(Crowther and Raymann-Bacchus, 2004; Hopkins, 2003). CSR has accumulated its own literature, typically discussed in organizational arenas, which tend to focus on larger corporations and the notion of 'corporate citizenship', measurement strategies, and balance of stakeholders' rights.

Possibilities of Leading Family Sustainability in Organizations

A study of two small business owners from group community relations in Illinois looked at their respective adherence to philosophies of CSR. These were micro-enterprises ranging in size from 1 to 10 employees, mostly in customer service, committed to social, environmental, and family sustainability. One group in Decatur, Illinois and one in Springfield, Illinois. Participants talked together about their meanings, challenges and strategies in developing practices of family sustainability while surviving as small businesses. Where possible, the businesses were also interconnected through in-depth interviews with the local media and civic leaders. While the analysis in these two studies proceeded independently according to their own unique issues and strategies, certain similarities became evident. Among the small business owners and community were strong commitments to social edification and corporate responsibility. Leaders emphasized the importance of facilitating citizen and employees regarding family sustainability. They also talked explicitly about constantly leading through everyday actions in different aspects of organizational operations.

CSR is promoted through many networks and alliances among business, community groups, trade unions, colleges and environmental activists. A wide range of instruments now express CSR goals, measure specific benchmarks and performance standards providing transparent inspection/assessment (e.g ILO, 2006; Verite, 2006). Willard (2005) claims that even executives continue to respond first to 'shareholder' demands, that since the mid-1990s they are responding to powerful and urgent demands converging from consumers, activist shareholders, nongovernmental organizations and governments alike. So, what are some family sustainability responses?

One seven year-old girl reported that reuniting with her father was "The best day of my life." Another child self-reported in her record "Love, love" while practicing family sustainability at a community room. Many other such emotional expressions have been formally documented on social media and in legal situations like mediation. Even though emotions and family sustainability feelings are not typically associated with processing the family disassembly through divorce and other such operations.

An analysis of organizational leaders like educators demonstrating CSR (corporate social responsibility) to pivot solutions in place of problems typically incubate six elements besides emotional learning. Those elements look at various degrees of decentralization, diversity/inclusion, connections, shared focus, relevant constraints, and feedback. These six elements are classically thought of as the six elements for learning. The elements that complex sciences have shown are characterized by complex adaptive systems. That is, systems that self-organize and increase rates of productivity because they are continuously adaptive and innovative. While emotions and family sustainability are also a known a condition, or prerequisite to productivity (Wemple, 2019).

Small Businesses and Family Sustainability

When seven elements including emotional constraint where detailed through participatory observation, then organizational production could be measured. Over one year a CLU (Clean Law Union) study of the production of two small businesses dropped to nearly zero when family sustainability was glitchy during the footprints of divorce. One previous study ended in death when the business owner of Taylor Trucking took his own life. It seemed shameful that no leaders consider family sustainability. And the thought at that time was that a measurement to understand weather or not a family could sustain a divorce, a small business, and their bottoms line over time was nearly impossible.

During the following year in the study of the two previously mentioned businesses, emotions and family sustainability along with practicing footsteps towards new policy initiatives aimed at correcting family dissassemblies actually led to a spike in organization productivity. Leading to the concept of family sustainability. For example, one leader reported that he was working on giving up hundreds of clients and going out of business before his practice of family sustainability. But after shifting towards family sustainability practices, the plans changed and he worked not only on maintained hundreds of clients but added 23 more clients. A current study is underway to detail if this increased production associated with family suitability practices will be maintained or not. Which leads to a of how much productivity and social well-being are lost by virtually no organizational efforts in family sustainability practices?

Emergence of Family Sustainability Leadership

This approach draws from literature's in family, social-emotional learning and corporate social responsibility to describe the nature of family sustainability, the intents and approaches of organizations developing it, as well as their challenges. Case studies are drawn from small businesses and are examined to explore successful strategies of developing practices in family sustainability. These examples are analyzed from an organizational leadership perspective through management applications. While the additional emotional element is very self-evident. Like when a pet owner leaves for work and the pet stays inside the home, then the pet feels those disconnections and the pets even exhibit crying sounds. Sometimes, they even tear up the environment that they are left in. Or, like when a parent leaves a child even temporarily, then nuclear families often feel those disconnections. Family unsustainability, in large part, is known by its human emotions and any social recoils. Which often translates into a lack of productivity and similar social effects.

Several conditions prevent crops, families and organizations from growing. Unsustainable ecosystems may be have next-generation restraints, centralization of decisions, exclusion of those effected by decisions, disconnections, and the lack of feedback. The conditions around family unsustainability are virtually unknown. However, the emotional elements are very evident. When a pet owner leaves a house for work with the pet still inside the home, then those pets are known to feel that disconnection and they often cry. Sometimes, they even tear up the environment that they are left in. Or, when a parent leaves a child even temporarily. American Academy of Pediatrics President Dr. Coleen Kraft (2018) says that everything pediatricians stand for is protecting and promoting children's health. Primarily including maintaining nuclear family bonds. Family unsustainability, in large part, is known by its human emotions and any social recoil. Which often translates into a lack of productivity and similar social effects.

Dr. Kraft (2018) explains about the children of national policy making issues, "Separating children from their parents contradicts everything we stand for as pediatricians – protecting and promoting children's health. In fact, highly stressful experiences, like family separation, can cause irreparable harm…"

Complex science like life processes teach us that emergence requires the conditions of diversity, decentralized organization, redundancy (overlap among individuals created through shared focus, language or activity), frequent opportunities for informal interaction, focus, and feedback (Fenwick, 2001; Davis and Sumara, 2005). Conditions for emergence do not necessarily occur spontaneously, but with the consideration for the emotional element of unfit ecosystems, then prevention of emergence is clear.

Quality Management and Family Sustainability

Rusinko (2005) uses quality management (QM) as a bridge for sustainability in organizations. QM, she says, is particularly suited to aid managers in implementing sustainable practices. QM, environmental sustainability and family sustainability have common themes illustrated by other management researchers (e.g., Hart, 1995; Porter and Van der Linde, 1995). These include a long-run organizational view, as well as a focus on both economic and social well-being. In addition, successful family sustainability, environmental sustainability and QM all acknowledge the importance of continuous improvement and participation by and empowerment of all employees. To maximum the benefits of QM, environmental sustainability and family sustainability, according to Rusinko (2005) managers should adopt an integrated, multi-functional, and organization-wide approach like the Deming Cycle of committing to continuous improvements.

The Deming Cycle is a continuous improvement methodology with four stages: plan, do, study, act (Dean and Evans, 1994).

In the **plane stage (plan)**, a current situation is studied, data is gathered, and a plan is developed for action. Let's take a Caterpillar manufacturing line for example. The management wants to increase productivity which takes a huge amount of synchronized team effort. They notice a glitch is routinely causing a bottleneck in products being built as the products move between various stations. During the planning stage, they notice that one employee has low morale, is routinely late for work, and others avoid that situation. Instead of directly replacing all human glitches and absorb those associated costs, a plan starts with just one adjustment to test the success.

The **second stage (do)** consists of implementing the plan on a trial or limited time basis. The old and new positions are tested in practice. The work is checked for one month. At the same time, however, the glitchy position continues to cause a bottleneck in the assembly line. This gives those responsible the advantage that they can now see exactly whether the adjustment was a human cause or something else. Although the production error has been contained, the production speed has hardly increased at all.

In the **third phase (study)**, the number and difficulty of parts to be installed at the slow station are counted and compared to the number and difficulty of other parts to be installed at various other stations. The study stage determines whether the trial plan is working and investigates any additional problems or opportunities. A re-adjustment is made equally distributing parts to be installed amount equally stations.

In the **act stage (act)**, the final plan (as tweaked in the study stage) can be fully implemented. Since all stations have the same number of parts and employee morale is elevated, the employees received instructions on new operating procedures. Since the Deming Cycle focuses on continuous improvement, the improvements that result from this final plan inspire further improvements and a return to the plan stage -and the rest of the cycle. QM bridges the gaps between costly human concerns and operation issues Rusinko (2005). While the Deming Cycle is widely used to help innovative strategies for continuous solutions.

Leading in the 21st Century

Clearly many obstacles can compromise efforts to incorporate practices of family sustainability in today's organizations trying to survive and compete in global markets, sufficient examples exist to inspire leaders and organizational developers to continue pressing for change. However, improving the emotional elements when a pet owner leaves for work or when a parent leaves their child would have valuable support. Improving the glitches associated with divorce and business often means the difference between production and no production when ignored. But the difference between production and super production when sustained.

Ultimately what matters is how individuals learn to resonate with the people and elements surrounding them with a greater sense of purpose, connection and mutual responsibility (Davidson and Hatt, 2005) This chapter began by showing that organizational practices of family sustainability, as explained in both academic literature and corporate social responsibility, can be represented by the four themes of ethical responsibility, renewal, interconnectivity and local well-being.

Albert Szent-Györgyi, who discovered Vitamin C, says that "Innovation is seeing what everyone else has seen and thinking what no one else has thought." Continuous family sustainability throughout the organization was a theme in the organizational examples described earlier. The strategies used by leaders to foster this ability are remarkably consistent with ecological models of learning derived from complexity science, the study of adaptive, self-organizing life systems. The focus is emergence: how are practices of family sustainability being developed, what management styles are involved, and how might these perspectives be fostered by organizational leaders.

Leaders can prompt frequent, informal interaction across different lines of an organization by modeling its importance, creating occasions for it, and posing questions mobilizing learners to gather issues and strategies. Leaders also can promote decentralized organization: different connections of control that communicate and cooperate. Effective leaders already tend to promote feedback within a system, feedback that attends a group to healthy directions and to negative loops that threaten to kill a system. The more loops for feedback that are created, according to complexity theory, the more that parts of an organization become attuned and interconnected with one other, with their central purposes, and with external environments and organizations.

References

AAP Statement Opposing Separation of Children and Parents at the Border. (n.d.). Retrieved July 6, 2019, from https://www.aap.org/en-us/about-the-aap/aap-pressroom/PagesStatement OpposingSeparationofChildrenandParents.aspx.

BSR (Business for Social Responsibility), (2006), Issue briefs: Business ethics, economic development and community investment, environment, governance and accountability, human rights, marketplace, workplace, retrieved July 4, 2019 from http://www.bsr.org.

Crowther, D. and Rayman-Bacchus. L. (Eds.), (2004), Perspectives on Corporate Social Responsibility, Ashgate, Aldershot, UK.

Daly, H.E. and Cobb Jr., J. (1989), For the Common Good: Redirecting the Economy toward Community, the Environment and a Sustainable Future, Beacon Press, Boston.

Davis, B., Sumara, D., and Luce-Kapler, R (2000), Engaging Minds: Learning and Teaching in a Complex World, Laurence Erlbaum, Mahwah, NJ.

Davidson, D.J. and Hatt, K.C. (2005), "Towards a sustainable future", in Davidson, D. and Hatt, K. C. (Eds.), Consuming Sustainability: Critical Social Analyses of Ecological Change, Winnipeg, Manitoba, pp. 228-244.

Dean, J.W., Jr., and Evans, J.R. (1994). Total quality management, organization, and strategy. St. Paul, MN: West Publishing.

Fenwick, T. (2001), Work knowing on the fly: Post-corporate enterprise cultures and co-emergent epistemology. Studies in Continuing Education, 23 (1), 243- 259.

Hart, S.L. (1995). A natural-resource-based view of the firm. Academy of Management Review, 20(4), 986-1014.

Hopkins, M. (2003), The Planetary Bargain: Corporate Social Responsibility Matters, Earthscan, London, UK.

ILO (International Labour Organisation), (2006), Tripartite Declaration of rinciples concerning Multinational Enterprises and Social Policy, retrieved on July 4, 2019 from http://www-ilo-irror.cornell.edu/Public/english/standards/norms/sources/mne.htm.

Kraft, C., MD. (n.d.). AAP Statement Opposing Separation of Children and Parents at the Border. Retrieved July 7, 2019, from https://www.aap.org/en-us/about-the-aap/aap-press-room/Pages/StatementOpposingSeparationofChildrenandPArents.aspx

Porter, M.E., and van der Linde, C. (1995). Green and competitive: Ending the stalemate. Harvard Business Review, 73(5), 120-34.

Rusinko, C. A. (2005). Using quality management as a bridge to environmental sustainability in organizations. *SAM Advanced Management Journal, 70*(4), 54.

Todd, N.J. and Todd, J. (1994), From Eco-Cities to Living Machines: Principles of Ecological Design, North Atlantic Books, Berkeley.

Verité. (2006), Social Accountability International, retrieved on July 4, 2019 from http://www.verite.org/network/frameset.htm.

www.ingramcontent.com/pod-product-compliance
Lightning Source LLC
Chambersburg PA
CBHW020525030426
42337CB00011B/552